ABCs of Poultry Science

Written and illustrated by
YUGUO TOMPKINS

Copyright © 2023 Yuguo Tompkins, US
All right reserved

Editor: Robert W. Tompkins

ISBN 979-8-9896421-0-6

Written permission must be secured from the publisher to use or reproduce any part of this book.

To Rose and Keai, thank you for being my inspiration and motivation.

To my husband, Philip Tompkins, and my parents, thank you for being my strength and support.

Grateful for the unwavering support of my family and friends.

A is for Avian

Our friends are so sweet,
with feathers and beaks,
their language is tweet.
From songbirds to penguins,
they're all so unique!

B is for Beak

Which birds use to eat,
they peck at their food,
it's such a treat!

C is for Chick

so small and so sleek,
tiny and fluffy,
a chirping mystique.
In barns it parades,
a feathery delight,
Endearing and small,
stealing hearts at first sight.

 is for Duck

they love swimming
with glee.
In a pond or a river,
they're as happy as
they can be.

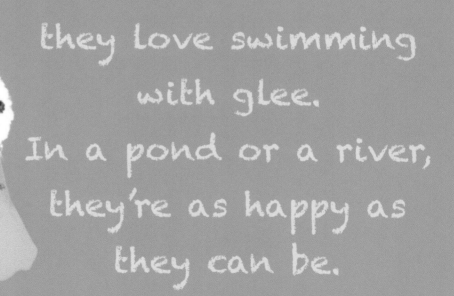

E is for Egg

life circle's start.
A shell round and smooth,
a work of art.

F is for Feather

so fluffy and fine;

a warm and cozy coat;

in which birds look divine.

G is for Goose

sleek feathers to greet.
From meat to down,
a poultry industry feat.
In the wild
and on the farm,
their role's truly sweet.

H is for Hatching

chicks start to peep, peck by peck,
breaking free from the sleep.
Twenty-one days, the magic unfolds,
tiny wings flutter, stories to be told.

I is for Incubator

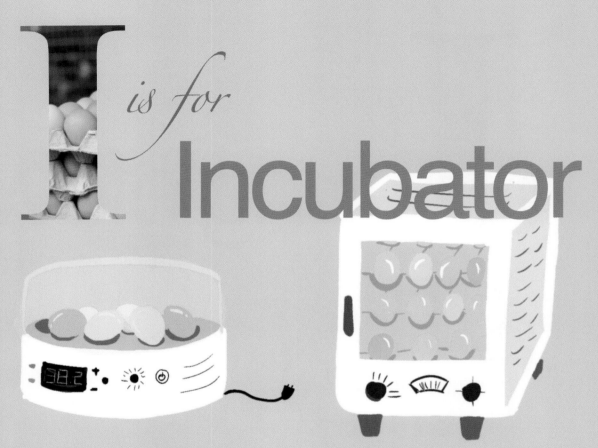

to keep the egg warm and secure;
a small home for baby birds to grow
and endure.

J is for Junglefowl

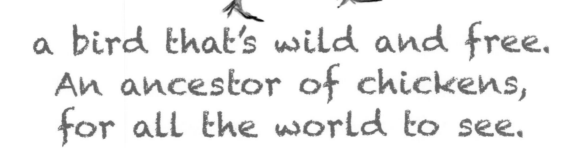

a bird that's wild and free.
An ancestor of chickens,
for all the world to see.

K is for Keel bone

where flight muscles find their base, injuries can happen behavior shows the trace. A good keel bone is essential, for happy hens to embrace.

L is for Layer

Who delivers eggs with care.
In its cozy nest, it lays eggs here and there!

M is for Molting

a cycle in the hen,
feathers fall off,
and new growth
does begin.

N is for Nest

Where birds create their best.
With twigs, grass and feathers,
they build their cozy nest.

O is for Ostrich

grand and tall,
majestic ruler,
reigning over all.
Swift and speedy,
legs so strong,
eggs in the ground
where they belong.

P is for Poultry

a value so grand,
feathers and eggs,
in great demand.
Quacking and clucking,
a lively band!

Small, shy, and cute.
for genetic studies,
they aid in scientific pursuit.

R is for Rooster

a crown on its head.
Vibrant feathers,
strutting the farmstead.
Cock-a-doodle-doo,
for me and you!

S is for Shell gland

where Calcium comes in waves, crafting a coat for eggs, as new life achieves.

T is for Turkey

a feathered delight,
gobbling and strutting
a festive sight.
With a wobble and gobble,
they make us smile.
A symbol of gratitude,
each year for a while.

U is for Uropygial gland

a feather's best friend.
Magic oil at
the tail's end,
on which the
bird can depend.

V is for Vet (Veterinarian)

with care so sweet, keeping birds healthy, from head to feet!

W is for Welfare

the care of all
our kin.
For happy birds
and happy
farmers,
their welfare is a
win!

X is for X-ray

to see the bones inside.
A healthy poultry structure
is what we aim to provide.

Y is for Yolk

a sunny shade
so bold.
It's nutritious
for us all,
both young and
growing old.

Z is for Zeal

a scientist's passion. Aiming for health, in a dedicated fashion. Providing safe food, their noble pursuit, ensuring animal welfare, their mission absolute.

Printed in Great Britain
by Amazon